Contents

Preface	1
Tuberculosis	3
isoniazid	14
pyridoxine	15
rifampicin	16
rifampicin/isoniazid	17
pyrazinamide	18
streptomycin	19
ethambutol	20
thioacetazone/isoniazid	21
BCG vaccine (dried)	22
tuberculin (purified protein derivative)	23
Leprosy	24
clofazimine	28
dapsone	29
rifampicin	30
ethionamide and protionamide	31
Nonspecific mycobacterial diseases	33
rifampicin	37
ethambutol	38
isoniazid	39

WHO Model Prescribing Information

Drugs used in Mycobacterial Diseases

World Health Organization
Geneva
1991

WHO Library Cataloguing in Publication Data
WHO model prescribing information : drugs used in mycobacterial diseases.
1.Antitubercular agents 2.Leprostatic agents 3.Mycobacterial infections - drug therapy 4.Prescriptions, Drug

ISBN 92 4 140103 6   (NLM Classification: QV 268)

© **World Health Organization 1991**

Publications of the World Health Organization enjoy copyright protection in accordance with the provisions of Protocol 2 of the Universal Copyright Convention. For rights of reproduction or translation of WHO publications, in part or *in toto*, application should be made to the Office of Publications, World Health Organization, Geneva, Switzerland. The World Health Organization welcomes such applications.

The designations employed and the presentation of the material in this publication do not imply the expression of any opinion whatsoever on the part of the Secretariat of the World Health Organization concerning the legal status of any country, territory, city or area or of its authorities, or concerning the delimitation of its frontiers or boundaries.

The mention of specific companies or of certain manufacturers' products does not imply that they are endorsed or recommended by the World Health Organization in preference to others of a similar nature that are not mentioned. Errors and omissions excepted, the names of proprietary products are distinguished by initial capital letters.

Printed in England
90/8715– PHA/Demandus – 5500

Preface

WHO's revised drug strategy, as adopted in Resolution WHA39.27 of the Thirty-ninth World Health Assembly in 1986, calls for the preparation of model prescribing information which is being developed to complement WHO's Model List of Essential Drugs.[1] The objective is to provide source material for adaptation by national authorities, particularly in developing countries, that wish to develop national drug formularies, drug compendia and similar material.[2]

The information is to be regarded as illustrative rather than normative. It is appreciated that it is not possible to develop an information sheet on a specific drug that is appropriate to circumstances prevailing in each of WHO's Member States and that some countries have already formally adopted texts of their own that have a statutory connotation.

This volume has been reviewed by internationally accredited experts, by all members of WHO's Expert Advisory Panel on Drug Evaluation and by members of the WHO expert advisory panels on mycobacterial diseases. It has also been reviewed by certain nongovernmental organizations in official relations with WHO, including the International Federation of Pharmaceutical Manufacturers Associations, the International Pharmaceutical Federation, the International Union of Pharmacology and the World Federation of Proprietary Medicine Manufacturers.

---

[1] WHO Technical Report Series, No. 796, 1990 (*The use of essential drugs:* fourth report of the WHO Expert Committee).
[2] Already published: *WHO model prescribing information: drugs used in anaesthesia*, 1989; and *WHO model prescribing information: drugs used in parasitic diseases*, 1990.

Drug dosage

Most drug doses are given per kilogram of body weight or as fixed doses calculated for adults of 60 kg.

Storage conditions

Readers are referred to *The International Pharmacopoeia*, 3rd edition, vol. 1 & vol. 2 (Geneva, World Health Organization, 1979 & 1981) for definitions concerning containers for drugs.

Tuberculosis

Tuberculosis occurs throughout the world and remains an important cause of morbidity and mortality in many developing countries. Each year, it is estimated that 8 million new cases occur and that it causes 3 million deaths. Whereas its prevalence has declined considerably in the industrialized world within the last 50 years, it has re-emerged among patients infected with human immunodeficiency virus (HIV).

The disease develops in only a small proportion of individuals infected with the causative mycobacteria, *Mycobacterium tuberculosis*, *M. bovis* and, in a few instances, *M. africanum*. Most cases are caused by *M. tuberculosis* for which humans constitute the only significant reservoir. The bacteria are generally transmitted in airborne droplets dispersed by patients with infectious pulmonary or laryngeal disease, and children are at particularly high risk if there is a family contact. *M. bovis* is additionally transmitted in milk from diseased cows.

Primary infection occurs when phagocytic defences fail locally and bacilli proliferate in macrophages, usually in a discrete alveolar focus. This may give rise to mild fever and malaise, particularly when the regional lymph nodes are involved, but most cases are asymptomatic. When the immune system is competent, the infection is generally contained within 3 to 6 weeks. Only isolated granulomata then remain, usually in the lungs or the regional lymph nodes, but sometimes in a more distant focus.

Occasionally, the entire immune system is overwhelmed, particularly in infants and in individuals who are immunodeficient, malnourished or debilitated. Massive haematogenous spread of the infection results either in acute miliary tuberculosis or in tuberculous meningitis. Diagnosis of these conditions, which is rarely possible on clinical examination alone, is confirmed in the former case by characteristic radiographic changes and in the latter by the detection of typical changes in the chemistry of the cerebrospinal fluid.

A healed primary infection normally confers substantial immunity. However, dormant lesions may be reactivated by malnutrition, corticosteroid therapy or immunosuppressive drugs and diseases, including HIV infection. Post-primary tuberculosis may also result from exogenous reinfection where the disease is highly prevalent. Constitutional symptoms are prominent in such cases: malaise, loss of weight, low-grade fever and night sweats supervene as the disease progresses. In most cases, the major focus of infection is in the lungs, but it may be elsewhere if haematogenous or lymphatic spread has occurred. Lymph nodes, kidneys, bones, the genital tract, the brain and the meninges are particularly vulnerable and the typical granulomatous foci tend to caseate, undergo cavitation and fibrose. In the lungs, caseous lesions often liquefy and drain into the bronchial tree giving rise to persistent cough, copious sputum and episodes of haemoptysis of varying degree. Bleeding can be massive when a bronchial artery is eroded, and a risk of miliary tuberculosis arises whenever a caseous focus ruptures into the bloodstream.

Prevention

Vaccination

The incidence of tuberculosis falls wherever living standards are raised and pasteurization of milk is introduced. Where the disease remains highly prevalent, routine immunization of infants within the first year with BCG vaccine, derived from bacille Calmette–Guérin (an attenuated strain of *M. bovis*), is highly cost-effective. This has been estimated, in several settings, to reduce the incidence of meningeal and miliary tuberculosis in early childhood by 50 to 90%. However, estimates of its effectiveness in older children have differed greatly from region to region.

Chemoprophylaxis

Chemoprophylaxis with isoniazid can prevent the development of clinically apparent disease in the close contacts of those infected and in others at high risk. However, it occasionally induces hepatitis, particularly in patients over 35 years of age, and symptomless individuals are not readily persuaded to take the drug regularly. In most countries the limited resources available for control programmes are more effectively directed to case-finding and treatment. All members of the immediate family and other close contacts of smear-positive patients should be traced, tested and, if necessary, followed up for at least one year.

Tuberculin testing

The tuberculin test is used to demonstrate cell-mediated skin-reactivity to tuberculin. Where there is a low prevalence of tuberculosis and BCG vaccine is not routinely administered to children, annual testing for tuberculin conversion is still used in a few countries as a means of identifying new cases. It has also been used to establish the level of immunity produced by BCG vaccination. However, it is important to appreciate that the test has limited diagnostic value. It indicates whether a person has at some time reacted immunologically to mycobacteria, but provides no means of distinguishing between tuberculous and nontuberculous mycobacterial infection, between active and quiescent disease, or between acquired infection and seroconversion induced by BCG vaccination. A negative response normally indicates that a person has never previously been infected by mycobacteria, but anergy—resulting from HIV and other viral infections, malnutrition, and treatment with corticosteroids or immunosuppressive drugs—can result in false-negative reactions.

Several variants of the test have been described, but the Mantoux technique, in which a measured quantity of tuberculin is injected, should always be preferred. Other tests, including the multiple-puncture (Heaf) test, are difficult to interpret and are not recommended for general use.

The depression or inhibition of the tuberculin reaction among HIV-infected patients is of particular importance since it compromises the diagnosis of tuberculosis within a population at high risk of endogenous reactivation. The humoral response to mycobacterial antigens is apparently not related to the tuberculin skin-reactivity, and studies are in progress to determine whether a serological test might detect circulating antibodies to mycobacterial antigens in people who are anergic to the tuberculin test.

Chemotherapy

General principles

Effective chemotherapy rapidly reduces the population of viable bacilli and consequently reduces the risk of transmission. The treatment of smear-positive cases should therefore be accorded highest priority in governmental control programmes. Health workers should be trained to recognize the clinical features of pulmonary tuberculosis and, when a case is suspected, to send sputum to a laboratory with facilities for smear examination. When possible, sputum cultures should also be prepared, but treatment should not be delayed once a presumptive diagnosis has been made. In smear-negative cases, chest radiographs can also provide strong evidence of disease, but radiological screening is usually too costly to be undertaken on a community scale and efforts should always be made to obtain bacteriological confirmation.

Institutional care is necessary initially for patients who fail to respond to treatment and for those who are severely ill (particularly as a result of a serious complication such as massive haemoptysis or pyopneumothorax) or who are bedridden with severe paraparesis. Most other patients can be treated satisfactorily at home provided that they can be relied upon to take their drugs regularly. Corticosteroids have no place in the routine management of tuberculosis but, in a hospital setting, short courses are of value in treating patients with pericarditis, pleural effusions, tuberculous meningitis and acute miliary tuberculosis associated with dyspnoea. Surgical resection is now necessary only in rare cases of severe post-tuberculous bronchiectasis, particularly when this is a cause of repeated haemoptysis.

Five drugs are currently regarded as essential in the management of tuberculosis—isoniazid, rifampicin, pyrazinamide, streptomycin and ethambutol. Thioacetazone is also widely used to supplement isoniazid in many developing countries because of its low cost. Other drugs, including *para*-aminosalicylic acid, kanamycin, cycloserine, capreomycin, viomycin and ethionamide, can be of value in treating patients with multiple-drug-resistant disease but, in general, they are more expensive and more toxic.

The management of patients with tuberculosis is challenging in several respects:

- Tubercle bacilli are vulnerable to bactericidal drugs only when they are metabolically active and replicating.

- Small subpopulations of bacilli remain semidormant indefinitely and become transiently active—and thus vulnerable to bactericidal agents—only for very short periods of time.

- Drug-resistant mutants can exist even in populations of bacilli never previously exposed to antibiotics.

An ideal antituberculosis drug, apart from being cheap and of low toxicity, would consequently need to possess:
- potent bactericidal activity against metabolically active bacilli;
- potent sterilizing activity against semidormant persisting bacilli; and
- potential to prevent the emergence of resistant organisms throughout the period of chemotherapy.

No known chemotherapeutic agent possesses each of these properties to a degree that renders it self-sufficient for the treatment of smear-positive cases. At least three drugs must be administered concomitantly for the first 2 months, in particular to prevent the emergence of resistance, and in all cases treatment must be continued with at least two drugs for several months longer before there can be reasonable assurance of cure.

Properties of antituberculosis drugs
The principles of antituberculosis therapy as set out above also provide a mechanistic basis for classifying the available drugs.

Bactericidal action
Isoniazid, which decimates the bacillary population within a few days of starting treatment, is unrivalled in its bactericidal potency. Rifampicin and ethambutol are moderately active, while streptomycin, pyrazinamide and thioacetazone have little or no bactericidal action. The intensity of the bactericidal effect determines the rate at which viable tubercle bacilli disappear from the sputum during the first few days of chemotherapy. It has little influence, however, on the speed at which less vulnerable bacilli are subsequently eliminated.

Sterilizing action
The sterilizing activity of a treatment regimen determines the required duration of therapy. Rifampicin and pyrazinamide are particularly effective in eliminating subpopulations of bacilli that are unresponsive to other drugs. Rifampicin is active against semidormant or intermittently active bacilli because it has the potential to kill these organisms during the short intervals when

they are vulnerable to chemotherapeutic attack. Because it is active in an acidic environment, pyrazinamide is effective uniquely in zones of acute inflammation and against quiescent bacilli within macrophages. It is not very effective during the latter phase of treatment after the acute inflammatory response has subsided.

Inhibitors of acquired resistance
Isoniazid is effective in suppressing the proliferation of mutants resistant to other drugs, as is rifampicin, since the inhibitory effect of both is adequately sustained even when occasional doses are taken irregularly. Streptomycin and ethambutol are less effective in this respect, while pyrazinamide and thioacetazone are the least effective. Despite this disadvantage, thioacetazone is commonly relied upon to inhibit the emergence of isoniazid-resistant strains in patients on long-term regimens, during the period when they are essentially reliant upon isoniazid. Combination tablets containing isoniazid and thioacetazone are consequently still used where no other treatment regimen can be afforded.

Preferred treatment regimens

National chemotherapy policies need to be reviewed periodically since advances continue to be made in the treatment of tuberculosis. The higher costs of the more potent drugs should always be set against the savings that accrue from shorter treatment regimens and a lower incidence of therapeutic failure and relapse. Regimens in which isoniazid and rifampicin are taken throughout can usually be relied upon to cure bacteriologically confirmed disease within 6 months. If rifampicin is taken only during the first phase of therapy the period of treatment must be extended by at least 2 months. Regimens that do not include either rifampicin or pyrazinamide must be sustained for at least 12 months to ensure elimination of persistent organisms.

Regimens recommended by WHO on the basis of extensive field trials are detailed on pages 12 and 13. Each has produced cure rates approaching 100% when drug-taking has been fully supervised, and one of them should be selected for all newly confirmed cases of tuberculosis.

The 6-month regimens of daily isoniazid and rifampicin, supplemented by pyrazinamide for the first 2 months, are regarded as the most effective. These drugs should always be available for use in patients with tuberculous meningitis because they all penetrate freely into the cerebrospinal fluid when the meninges are inflamed. If resistance to any one of these three drugs is suspected, ethambutol or streptomycin should be added for the first 2 months. Ethambutol is of particular value as an ancillary drug in industrialized countries, where the patients tend to be older and adequate ocular examinations can usually be carried out to minimize the risk of dose-dependent optic neuritis. Streptomycin may be preferred in developing countries, where the patients are generally younger (and therefore at lower risk of vestibular damage due to streptomycin), and where it is more difficult to screen adequately for optic neuritis. Moreover, streptomycin helps to ensure compliance because its administration must be supervised. However, it is necessary to exclude all possibility of transmission of hepatitis viruses and HIV infection by using a sterile syringe and needle for each patient. Streptomycin should never be given during pregnancy because it can readily cause permanent auditory and vestibular damage in the fetus.

Where adequate facilities for fully supervised administration exist, a modified regimen in which the same drugs are taken three times weekly can be used.

An 8-month regimen, comprising supervised administration of isoniazid, rifampicin, pyrazinamide and either ethambutol or streptomycin for 2 months, followed by 6 months' self admin istration of isoniazid and thioacetazone, is also effective and is used in some countries to reduce drug costs.

The 12-month regimens are much less effective and should be used only when the more potent drugs are not available.

Monitoring the therapeutic response

Inadequate compliance is by far the most important cause of therapeutic failure. The need to adhere rigorously to treatment schedules must be emphasized repeatedly to every patient. Whenever possible, drug administration should be supervised for at least the first 2 months. When professional supervision is not possible, another member of the household or a volunteer in the community should be taught to assume this responsibility.

Where the facilities for outpatient care are suitably developed, full supervision of treatment holds advantage and, in these circumstances, the intermittent regimens are particularly cost-effective. Tablets containing isoniazid and rifampicin, or isoniazid, rifampicin and pyrazinamide in fixed dosage combination hold advantage over single drugs in that they promote compliance and prevent the development of resistance. They should be used, however, only if acceptable bioavailability of each of the active constituents has been adequately assured. Treatment failures have resulted from poor manufacture of these combination products in some countries.

Response to treatment is most readily evaluated from the clinical response of the patient and particularly by monitoring weight changes, fever and cough. However, whenever resources permit, cultures and smears should be examined systematically. Nonculturable (and hence noninfective) bacilli occasionally persist in the sputum for several months after successful treatment. None the less, smear examination remains a useful aid to compliance. If possible, paired sputum samples should be examined after 2 months and after 5 months of treatment. If any one of these four samples is positive, two further pairs taken 4 weeks apart should be examined. If bacilli are again detected, every effort should be made to determine whether the patient is taking the drugs as prescribed. If noncompliance can be excluded, the patient should be transferred to another regimen.

Treatment of relapsing and unresponsive disease

Most cases of chronic or relapsing tuberculosis are the result of irregular, inadequate administration of prescribed drugs or of the less potent regimens used in the past. Whenever possible, the drug sensitivity of the bacilli should be established, and previous treatment should always be reviewed to determine whether resistance has developed and, if so, how it is likely to have happened. In the absence of reliable sensitivity tests, treatment with isoniazid, rifampicin and ethambutol should be instituted for 8 months, supplemented with pyrazinamide for the first 3 months and streptomycin for the first 2 months. All drugs should be administered under strict supervision for at least the first 2 months, preferably in a hospital setting.

Drugs used in pregnancy

Treatment should never be interrupted or postponed during pregnancy or at any other time when immunological resistance to the disease is reduced. However, the fetus should never be exposed to streptomycin because of the risk of auditory and vestibular nerve damage. Since, as a general principle, as few drugs as possible should be administered during pregnancy and for the shortest possible time, the 6-month regimen of daily rifampicin and isoniazid supplemented with pyrazinamide for the first 2 months holds advantage.

Promising developments

Antimycobacterial activity has been demonstrated in various newly developed antimicrobial substances. The 4-fluoroquinolones are promising bactericidal agents. Other compounds of interest include rifamycin derivatives with long half-lives. These, it is hoped, may provide a basis for fully supervised intermittent regimens. Combinations of clavulanic acid and ß-lactam antibiotics are also undergoing clinical assessment.

Tuberculosis and HIV infection

Because of immunosuppression, patients with HIV infection are at high risk of developing clinical tuberculosis. The disease, which is frequently atypical and extrapulmonary, often antedates manifestations of AIDS by months or even years. Standard 6- or 8-month regimens are usually curative provided that treatment is continued for at least 3 months after sputum conversion. A recommendation on whether subsequent chemoprophylaxis should be offered routinely to HIV-infected individuals will depend upon the results of ongoing trials involving, among others, regimens based upon isoniazid alone, rifampicin alone and combinations of rifampicin and pyrazinamide.

Although concern has been expressed about disseminated disease resulting from BCG vaccination in HIV-positive individuals, very few cases have been reported. The balance of evidence suggests that, in countries where the risk of tuberculosis infection remains high, every infant without signs of AIDS should be vaccinated, even when the mother is HIV-positive.

Recommended treatment regimens for tuberculosis[a]

Drug	6-month regimens	
	Phase 1: 2 months	Phase 2: 4 months
Isoniazid	5 mg/kg daily	5 mg/kg daily
Rifampicin	10 mg/kg daily	10 mg/kg daily
Pyrazinamide	30 mg/kg daily	
supplemented, in areas where resistance to one of these drugs is demonstrated, by		
Streptomycin *or*	15 mg/kg daily	
Ethambutol	25 mg/kg daily[b]	
OR		
Isoniazid	15 mg/kg 3 times weekly	15 mg/kg 3 times weekly
Rifampicin	10 mg/kg 3 times weekly	10 mg/kg 3 times weekly
Pyrazinamide	50 mg/kg 3 times weekly	50 mg/kg 3 times weekly
together with		
Streptomycin *or*	15 mg/kg 3 times weekly	15 mg/kg 3 times weekly
Ethambutol	40 mg/kg 3 times weekly[c]	40 mg/kg 3 times weekly[c]

Drug	8-month regimens	
	Phase 1: 2 months	Phase 2: 6 months
Isoniazid	5 mg/kg daily	5 mg/kg daily
Rifampicin	10 mg/kg daily	
Pyrazinamide	30 mg/kg daily	
Thioacetazone		2.5 mg/kg daily
together with		
Streptomycin *or*	15 mg/kg daily	
Ethambutol	25 mg/kg daily[b]	

[a] Unless otherwise indicated, doses are suitable for both adults and children.
[b] 15 mg/kg for children.
[c] Not suitable for children.

Drug	12-month regimens[a]	
	Phase 1: 2 months	Phase 2: 10 months
Isoniazid	5 mg/kg daily	5 mg/kg/daily
Streptomycin	15 mg/kg daily	
together with		
Thioacetazone	2.5 mg/kg daily	2.5 mg/kg daily
or		
Ethambutol	25 mg/kg daily[b]	15 mg/kg daily
OR		
Isoniazid	5 mg/kg daily	15 mg/kg twice weekly
Streptomycin	15 mg/kg daily	15 mg/kg twice weekly
Thioacetazone	2.5 mg/kg daily	

[a] Unless otherwise indicated, doses are suitable for both adults and children.
[b] 15 mg/kg for children.

Isoniazid

Group: antimycobacterial agent
Tablet 100–300 mg
Injection 25 mg/ml in 2-ml ampoule

General information

Isoniazid, the hydrazide of isonicotinic acid, is highly bactericidal against replicating tubercle bacilli.

It is rapidly absorbed and diffuses readily into all fluids and tissues. The plasma half-life, which is genetically determined, varies from less than 1 hour in fast acetylators to more than 3 hours in slow acetylators. It is largely excreted in the urine within 24 hours, mostly as inactive metabolites.

Clinical information

Uses

A component of all antituberculosis chemotherapeutic regimens currently recommended by WHO (see pages 12 and 13).

Isoniazid alone is occasionally used prophylactically to prevent:

- transmission to close contacts at high risk of disease, and
- recrudescence of infection in immunodeficient individuals.

Dosage and administration

Isoniazid is normally taken orally but it may be administered intramuscularly to critically ill patients.

Treatment (combination therapy)

Adults and children: 5 mg/kg (maximum 300 mg) daily or 15 mg/kg two or three times weekly.

Prophylaxis

Adults: 300 mg daily for 6 months to 1 year.

Children: 10 mg/kg (maximum 300 mg) daily for 6 months to 1 year.

Contraindications

- Known hypersensitivity.
- Active hepatic disease.

Precautions

Serum concentrations of hepatic transaminases should be monitored whenever possible.

Patients at risk of peripheral neuropathy as a result of malnutrition, chronic alcohol dependence or diabetes should additionally receive pyridoxine, 10 mg daily. Where the standard of health in the community is low this should be offered routinely.

Epilepsy should be effectively controlled since isoniazid may provoke attacks.

Use in pregnancy

Whenever possible, the 6-month regimen based upon isoniazid, rifampicin and pyrazinamide should be used.

Adverse effects

Isoniazid is generally well tolerated at recommended doses. Systemic or cutaneous hypersensitivity reactions occasionally occur during the first weeks of treatment.

The risk of peripheral neuropathy is excluded if vulnerable patients receive daily supplements of pyridoxine. Other less common forms of neurological disturbance, including optic neuritis, toxic psychosis and generalized convulsions, can develop in susceptible individuals, particularly in the later stages of treatment, and occasionally necessitate the withdrawal of isoniazid.

Hepatitis is an uncommon but potentially serious reaction that can usually be averted by prompt withdrawal of treatment. More often, however, a sharp rise in serum concentrations of hepatic transaminases at the outset of treatment is not of clinical significance. If the enzyme levels drop rapidly when dosage is suspended, they are unlikely to rise sharply again when treatment is resumed.

Drug interactions
Isoniazid tends to raise plasma concentrations of phenytoin and carbamazepine by inhibiting their metabolism in the liver. The absorption of isoniazid is impaired by aluminium hydroxide.

Overdosage
Nausea, vomiting, dizziness, blurred vision and slurring of speech occur within 30 minutes to 3 hours of overdosage. Massive poisoning results in coma preceded by respiratory depression and stupor. Severe intractable seizures may occur. Emesis and gastric lavage can be of value if instituted within a few hours of ingestion. Subsequently, haemodialysis may be of value. Administration of high doses of pyridoxine is necessary to prevent peripheral neuritis.

Storage
Tablets should be kept in well-closed containers, protected from light. Solution for injection should be stored in ampoules protected from light.

Pyridoxine

Group: vitamin
Tablet 10 mg (hydrochloride)

General information
Pyridoxine is the naturally occurring form of vitamin B_6. Following absorption from the gastrointestinal tract, it is converted in the liver to a coenzyme, pyridoxal phosphate, that is involved in many metabolic processes. Isoniazid interferes competitively with pyridoxine metabolism by inhibiting the formation of the active form of the vitamin, and hence often results in peripheral neuropathy.

Clinical information
Uses
To prevent the development of peripheral neuropathy in patients receiving isoniazid.

Dosage and administration
Adults and children: 10 mg daily suffices for prophylaxis but patients with evidence of vitamin B_6 deficiency may require up to 50 mg daily.

Precautions
Pyridoxine can block the therapeutic effect of levodopa by enhancing its decarboxylation to dopamine, which does not enter the brain. However, it does not interfere with the combined preparations of levodopa and a decarboxylase inhibitor.

Storage
Pyridoxine tablets should be stored in well-closed containers.

Rifampicin

Group: antimycobacterial agent
Capsule or tablet 150 mg, 300 mg

General information

A semisynthetic derivative of rifamycin, a complex macrocyclic antibiotic that inhibits ribonucleic acid synthesis in a broad range of microbial pathogens. It has bactericidal action and a potent sterilizing effect against tubercle bacilli in both cellular and extracellular locations.

Rifampicin is lipid-soluble. Following oral administration, it is rapidly absorbed and distributed throughout the cellular tissues and body fluids; if the meninges are inflamed, significant amounts enter the cerebrospinal fluid. A single dose of 600 mg produces a peak serum concentration of about 10 micrograms/ml in 2–4 hours, which subsequently decays with a half-life of 2–3 hours. It is extensively recycled in the enterohepatic circulation, and metabolites formed by deacetylation in the liver are eventually excreted in the faeces.

Since resistance readily develops, rifampicin must always be administered in combination with other effective antimycobacterial agents.

Clinical information

Uses

A component of all 6- and 8-month antituberculosis chemotherapeutic regimens currently recommended by WHO (see pages 12 and 13).

Dosage and administration

Rifampicin should preferably be given at least 30 minutes before meals, since absorption is reduced when it is taken with food.

Adults and children: 10 mg/kg (maximum 600 mg) daily or three times weekly.

Contraindications

- Known hypersensitivity to rifamycins.
- Hepatic dysfunction.

Precautions

Serious immunological reactions resulting in renal impairment, haemolysis or thrombocytopenia are on record in patients who resume taking rifampicin after a prolonged lapse of treatment. In this rare situation it should be immediately and definitively withdrawn.

Careful monitoring of liver function is required in the elderly and in patients who are alcohol-dependent or have hepatic disease.

Patients should be warned that treatment may produce reddish coloration of urine, tears, saliva and sputum, and that contact lenses may be irreversibly stained.

Use in pregnancy

Whenever possible, the 6-month regimen based upon isoniazid, rifampicin and pyrazinamide should be used.

Vitamin K should be administered to the infant at birth because of the risk of postnatal haemorrhage.

Adverse effects

Rifampicin is well tolerated by most patients at currently recommended doses, although gastrointestinal intolerance can be unacceptably severe. Other adverse effects (skin rashes, fever, influenza-like syndrome and thrombocytopenia) are more likely to occur with intermittent administration. Temporary oliguria, dyspnoea and haemolytic anaemia have also been reported in patients taking the drug three times weekly. These reactions usually subside if the regimen is changed to one with daily dosage.

Moderate rises in serum concentrations of bilirubin and transaminases, which are common at the outset of treatment, are often transient and without clinical significance. However, dose-related hepatitis can occur, which is potentially fatal. It is consequently important not to exceed the maximum recommended daily dose of 10 mg/kg (600 mg).

Drug interactions
Rifampicin induces hepatic enzymes, and may increase the dosage requirements of drugs metabolized in the liver. These include corticosteroids, steroid contraceptives, oral hypoglycaemic agents, oral anticoagulants, phenytoin, cimetidine, quinidine, ciclosporin and digitalis glycosides. Patients should consequently be advised to use a nonhormonal method of birth control throughout treatment and for at least 1 month subsequently.

Biliary excretion of radiocontrast media and sulfobromophthalein sodium may be reduced and microbiological assays for folic acid and vitamin B_{12} disturbed.

Overdosage
Gastric lavage may be of value if undertaken within a few hours of ingestion. Very large doses may depress central nervous function. There is no specific antidote and treatment is supportive.

Storage
Capsules and tablets should be kept in tightly closed containers, protected from light.

Rifampicin/isoniazid

Group: antituberculosis agent
Tablet 150 mg + 100 mg, 300 mg + 150 mg

General information
A fixed combination of rifampicin and isoniazid has been developed as an aid to compliance. It is essential that all such products are shown to have adequate bioavailability.

Clinical information

Uses
Both drugs are components of all 6- and 8-month antituberculosis chemotherapeutic regimens currently recommended by WHO (see pages 12 and 13).

Dosage and administration
Adults:
> 50 kg: 2 tablets (300 mg rifampicin + 150 mg isoniazid) daily.
< 50 kg: 3 tablets (150 mg rifampicin + 100 mg isoniazid) daily.

The combination tablets are not suitable for paediatric use.

For contraindications, precautions, use in pregnancy, adverse effects and drug interactions, see information given for the separate components.

Storage
Tablets should be stored in tightly closed containers.

Pyrazinamide

Group: antimycobacterial agent
Tablet 500 mg

General information

A synthetic analogue of nicotinamide that is only weakly bactericidal against *M. tuberculosis*, but has potent sterilizing activity, particularly in the relatively acidic intracellular environment of macrophages and in areas of acute inflammation. It is highly effective during the first 2 months of treatment while acute inflammatory changes persist and its use has enabled treatment regimens to be shortened and the risk of relapse to be reduced.

It is readily absorbed from the gastrointestinal tract and is rapidly distributed throughout all tissues and fluids. Peak plasma concentrations are attained in 2 hours and the plasma half-life is about 10 hours. It is metabolized mainly in the liver and is excreted largely in the urine.

Clinical information

Uses

A component of all 6- and 8-month antituberculosis chemotherapeutic regimens currently recommended by WHO (see pages 12 and 13).

Dosage and administration

Adults and children (for the first two months): 30 mg/kg daily or 50 mg/kg three times weekly.

Contraindications

- Known hypersensitivity.
- Severe hepatic impairment.

Precautions

Patients with diabetes should be carefully monitored since blood glucose concentrations may become labile. Gout may be exacerbated.

Use in pregnancy

Although the safety of pyrazinamide in pregnancy has not been established, the 6-month regimen based upon isoniazid, rifampicin and pyrazinamide should be used whenever possible.

Adverse effects

Pyrazinamide is usually well tolerated. Hypersensitivity reactions are rare, but some patients complain of slight flushing of the skin.

Moderate rises in serum transaminase concentrations are common during the early phases of treatment. Severe hepatotoxicity is rare.

As a result of inhibition of renal tubular secretion, a degree of hyperuricaemia usually occurs, but this is often asymptomatic. Gout requiring treatment with allopurinol occasionally develops. Arthralgia, particularly of the shoulders, commonly occurs and is responsive to simple analgesics. Both hyperuricaemia and arthralgia may be reduced by prescribing regimens with intermittent administration of pyrazinamide.

Overdosage

Little has been recorded on the management of pyrazinamide overdose. Acute liver damage and hyperuricaemia have been reported. Treatment is essentially symptomatic. Emesis and gastric lavage may be of value if undertaken within a few hours of ingestion. There is no specific antidote and treatment is supportive.

Storage

Tablets should be stored in tightly closed containers, protected from light.

Streptomycin

Group: antimycobacterial agent
Powder for injection 1 g base (as sulfate) in vial

General information

An aminoglycoside antibiotic derived from *Streptomyces griseus* that is used in the treatment of tuberculosis and sensitive Gram-negative infections.

Streptomycin is not absorbed from the gastrointestinal tract but, after intramuscular administration, it diffuses readily into the extracellular component of most body tissues and it attains bactericidal concentrations, particularly in tuberculous cavities. Little normally enters the cerebrospinal fluid, although penetration increases when the meninges are inflamed. The plasma half-life, which is normally 2–3 hours, is considerably extended in the newborn, in the elderly and in patients with severe renal impairment. It is excreted unchanged in the urine.

Clinical information

Uses
A component of several combined antituberculosis chemotherapeutic regimens currently recommended by WHO (see pages 12 and 13), of particular use when primary resistance to other drugs is suspected.

Dosage and administration
Streptomycin must be administered by deep intramuscular injection. Syringes and needles should be adequately sterilized, to exclude any risk of transmitting viral pathogens.

Adults and children: 15 mg/kg daily or two or three times weekly. Patients over 60 years may not be able to tolerate more than 500–750 mg daily.

Contraindications
- Known hypersensitivity.
- Auditory nerve impairment.
- Myasthenia gravis.

Precautions
Should hypersensitivity reactions occur, as is common during the first weeks of treatment, streptomycin should be withdrawn immediately. Once fever and skin rash have resolved, desensitization may be attempted.

Streptomycin should be avoided, when possible, in children because the injections are painful and irreversible auditory nerve damage may occur. Both the elderly and patients with renal impairment are also vulnerable to dose-related toxic effects resulting from accumulation. Serum levels should be monitored periodically and dosage adjusted appropriately to ensure that plasma concentrations, as measured when the next dose is due, do not rise above 4 micrograms/ml.

Protective gloves should be worn when streptomycin injections are administered, to avoid sensitization dermatitis.

Use in pregnancy
Streptomycin should not be used in pregnancy. It crosses the placenta and can cause auditory nerve impairment and nephrotoxicity in the fetus.

Adverse effects
Injections are painful and sterile abscesses can form at injection sites. Hypersensitivity reactions are common and can be severe.

Impairment of vestibular function is uncommon with currently recommended doses. Dosage should be reduced if head-

Streptomycin (continued)

ache, vomiting, vertigo and tinnitus occur.

Streptomycin is less nephrotoxic than other aminoglycoside antibiotics. None the less, close monitoring of renal function is necessary. Dosage must be reduced by half immediately if urinary output falls, if albuminuria occurs or if tubular casts are detected in the urine.

Haemolytic anaemia, aplastic anaemia, agranulocytosis, thrombocytopenia and lupoid reactions are rare adverse effects.

Drug interactions
Other ototoxic or nephrotoxic drugs should not be administered to patients receiving streptomycin. These include other aminoglycoside antibiotics, amphotericin B, cefalosporins, etacrynic acid, ciclosporin, cisplatin, furosemide and vancomycin.

Streptomycin may potentiate the effect of neuromuscular blocking agents administered during anaesthesia.

Overdosage
Haemodialysis can be beneficial. There is no specific antidote and treatment is supportive.

Storage
Solutions retain their potency for 48 hours after reconstitution at room temperature and for up to 14 days when refrigerated. Powder for injection should be stored in tightly closed containers protected from light.

Ethambutol

Group: antimycobacterial agent
Tablet 100–400 mg (hydrochloride)

General information

A synthetic congener of 1,2-ethanediamine that is bactericidal against *M. tuberculosis*, *M. bovis* and some nonspecific mycobacteria. It is used in combination with other antituberculosis drugs to prevent or delay the emergence of resistant strains.

It is readily absorbed from the gastrointestinal tract. Plasma concentrations peak in 2–4 hours and decay with a half-life of 3–4 hours. Ethambutol is excreted in the urine both unchanged and as inactive hepatic metabolites. About 20% is excreted in the faeces as unchanged drug.

Clinical information

Uses
An optional component of several combined antituberculosis chemotherapeutic regimens currently recommended by WHO (see pages 12 and 13), of particular use when primary resistance to other drugs is suspected.

Dosage and administration
Adults: 25 mg/kg daily for no more than 2 months followed, as appropriate, by 15 mg/kg daily; or 40 mg/kg three times weekly.

Children: 15 mg/kg daily.

Dosage must always be carefully calculated on a weight basis to avoid toxicity, and should be reduced in patients with impaired renal function.

Contraindications
- Known hypersensitivity.
- Pre-existing optic neuritis from any cause.

- Inability (for example due to young age) to report symptomatic visual disturbances.
- Creatinine clearance of less than 50 ml/minute.

Precautions
Patients should be advised to discontinue treatment immediately and to report to a doctor should their sight or perception of colour deteriorate. Patients who are too young or who are otherwise unable to comprehend this warning should not receive ethambutol.

Whenever possible, renal function should be assessed before treatment.

Use in pregnancy
Where there is no evidence of primary resistance, the 6-month regimen based upon isoniazid, rifampicin and pyrazinamide should be used. If a fourth drug is needed during the initial phase, ethambutol should be preferred to streptomycin.

Adverse effects
Dose-dependent optic neuritis can readily result in impairment of visual acuity and colour vision. Early changes are usually reversible, but blindness can occur if treatment is not discontinued promptly.

Signs of peripheral neuritis occasionally develop in the legs.

Overdosage
Emesis and gastric lavage may be of value if undertaken within a few hours of ingestion. Subsequently, dialysis may be of value. There is no specific antidote and treatment is supportive.

Storage
Tablets should be stored in well-closed containers.

Thioacetazone/isoniazid

Group: antituberculosis agent
Tablet 50 mg + 100 mg, 150 mg + 300 mg

General information
A fixed combination of thioacetazone and isoniazid that is almost as cheap as isoniazid alone and is intended to promote compliance. Thioacetazone, a thiosemicarbazone that is bacteriostatic against *M. tuberculosis*, is used in antituberculosis chemotherapy to inhibit the emergence of resistance to isoniazid, particularly in the continuation phase of the long-term regimens. It is well absorbed from the gastrointestinal tract. Peak concentrations in plasma are attained after 4–6 hours and the plasma half-life is about 12 hours. About one-third of the oral dose is excreted in the urine unchanged. (For general information on isoniazid see page 14.)

Clinical information

Uses
A component of some of the longer antituberculosis chemotherapeutic regimens currently recommended by WHO (see pages 12 and 13).

Dosage and administration
Adults: 150 mg thioacetazone + 300 mg isoniazid daily.

Children: 50 mg thioacetazone + 100 mg isoniazid daily.

Contraindications
- Known hypersensitivity to either component.

Thioacetazone/isoniazid (continued)

Precautions
Treatment should be withdrawn immediately if a rash or other signs suggestive of hypersensitivity occur.

Adverse effects
Effects attributable to isoniazid are listed on page 14. The thioacetazone component frequently causes nausea, vomiting, diarrhoea and skin rashes.

Rare cases of fatal exfoliative dermatitis and acute hepatic failure have been reported. Cases of agranulocytosis, thrombocytopenia and aplastic anaemia are also on record.

Dose-related ototoxicity is rare, but particularly careful monitoring is required when thioacetazone is used in combination with streptomycin.

Overdosage
Emesis and gastric lavage may be of value if undertaken within a few hours of ingestion. There is no specific antidote and treatment is supportive.

Storage
Tablets should be kept in well-closed containers.

BCG vaccine (dried)

Group: vaccine
Intradermal injection

General information
Bacille Calmette–Guérin (BCG) vaccines are lyophilized preparations of a live, attenuated strain of *M. bovis*. Various strains, all derived from the same original culture, but varying in their DNA structure, immunogenictiy and biochemical characteristics, are used for the manufacture of BCG vaccines. When administered shortly after birth they may provide substantial protection against severe forms of tuberculosis of childhood, including miliary tuberculosis and tuberculous meningitis. The response has been less consistent when they have been administered to older children.

Clinical information

Uses
To confer active immunity against tuberculosis. In countries where the risk of infection is high, it is recommended that all infants, including those born to HIV-positive mothers, should be vaccinated during the first year of life, provided that they have no clinical signs suggestive of AIDS.

Dosage and administration
It is important that the vaccine be administered intradermally (not subcutaneously) to avoid risk of abscesses.

It must be reconstituted with an appropriate diluent immediately before injection.

Neonates and infants: 0.05 ml by intradermal injection in the arm over the insertion of the deltoid muscle.

Children over 1 year old: 0.1 ml administered as above.

Contraindications
- Generalized eczema.
- Hypogammaglobulinaemia and immunodeficiency resulting from treatment with antimetabolites, irradiation or systemic corticosteroids.

Precautions
If an infective dermatosis such as scabies is present, an injection site where there is no infection should be selected.

Adverse effects
Lymphadenitis, osteitis and localized necrotic ulceration may occur.

Very rarely, cases of disseminated BCG infection have been reported in immunodeficient patients. This condition is generally rapidly responsive to antituberculosis chemotherapy.

Storage
Lyophilized vaccines should be protected from light and the storage temperature should not be allowed to rise above 8 °C. BCG vaccine withstands prolonged freezing without deterioration. Unused portions of the reconstituted vaccine should be discarded at the end of the working day.

Tuberculin (purified protein derivative)
Group: diagnostic agent
Intradermal injection

General information
Sterile aqueous solutions for intradermal injection are prepared by prolonged cultivation of human strains of *M. tuberculosis*, heating to 100 °C to kill the bacilli and subsequent filtering. The active fraction, which is mainly protein, is isolated by precipitation.

Clinical information
Uses
A diagnostic agent for detecting cell-mediated skin-reactivity to tuberculin.

Dosage and administration
For the Mantoux test, a special tuberculin syringe is used to inject 0.1 ml of tuberculin (5 IU) intradermally into the flexor surface of the upper forearm after cleansing with acetone or ether. The needle is held parallel to the skin and downward pressure applied until the point penetrates the superficial layers of the skin. It is then moved forward to the intended site of injection where a weal about 7 mm in diameter is raised. The test, which is read after 48–72 hours, is regarded as positive if an area of more than 10 mm diameter around the injection site is indurated.

Precautions
Care should be taken to prevent contact with open cuts, abraded or diseased skin, the eyes or the mouth.

Storage
Tuberculin preparations should be stored protected from light at temperatures between 2 and 8 °C. Any unused portion should be discarded.

Leprosy

Leprosy is a chronic mycobacterial infection that affects some 10 to 12 million people, mainly in Africa, Asia and South America.

The causative organism, *Mycobacterium leprae*, is a slow-growing intracellular bacillus that infiltrates the skin, the peripheral nerves, the nasal and other mucosae and the eyes. The disease is transmitted directly from person to person when bacilli are shed from the nose and open skin lesions of patients harbouring large numbers of organisms. *M. leprae* can enter the body through skin abrasions but the respiratory tract is probably the main portal of entry. The leprosy bacilli presumably cross the pulmonary alveoli without causing a primary lesion and reach their sites of nidation by haemogenous spread. The household contacts of leprosy patients are at greatest risk of acquiring the disease. However, most individuals have considerable natural immunity and many infections are suppressed. Indeed, clinical leprosy can be regarded as a consequence of deficient cell-mediated immunity in susceptible individuals.

Paucibacillary leprosy[1] results when cellular immunity is only partially deficient. Relatively few bacilli are demonstrable in skin smears. Granulomatous lesions in the dermis, which occasionally heal spontaneously, present as hypopigmented and hypoaesthetic or anaesthetic patches. Peripheral nerve involvement may result in no more than minor, localized impairment of sensation but, in severe cases, extensive sensory and motor loss induces trophic changes, muscle wasting and contractures.

Multibacillary leprosy[2] occurs when cellular immunity is largely deficient. Rugose, nodular skin lesions result from infiltration of the dermis by incompetent macrophages loaded with *M. leprae*. Nerve damage also commonly occurs which, if left untreated, may lead to crippling deformities. This damage is mostly sus-

---

[1] Includes: smear-negative indeterminate and tuberculoid leprosy (Madrid classification); smear-negative indeterminate, polar tuberculoid, and borderline tuberculoid leprosy (Ridley and Jopling classification).
[2] Includes: lepromatous and borderline leprosy (Madrid classification); polar lepromatous, borderline lepromatous and mid-borderline leprosy (Ridley and Jopling classification); and smear-positive cases of other types of leprosy.

tained during immunologically mediated inflammatory exacerbations, which are of two types:

- Type I (reversal reactions) resulting from a cell-mediated immune process and characterized by acute exacerbation of skin lesions and by focal or more generalized attacks of neuritis, sometimes resulting in permanent nerve damage.
- Type II (erythema nodosum leprosum) resulting from an immune-complex reaction and characterized by an antibody-dependent response. Discrete acute inflammatory lesions develop in the skin. Systemic symptoms, when they occur, include fever, lymphadenopathy, acute iridocyclitis and, less frequently, neuritis, polyarthritis and glomerulonephritis.

Visual impairment or blindness is frequent in both paucibacillary and multibacillary leprosy. It results either from mycobacterial infiltration and inflammation of structures in the anterior segment of the eye, or from trophic changes following damage to the trigeminal and facial nerves, resulting in lagophthalmos, deformed eyelids or corneal anaesthesia.

Control

Effective control of leprosy depends upon:

- efficient case detection, case-holding and treatment, and
- surveillance of contacts.

Those at risk, and particularly close family contacts, should remain under periodic surveillance whenever possible.

Vaccines containing a suspension of killed *M. leprae* are currently being field-tested but none is yet available for routine use.

Chemotherapy

Dapsone has served as the mainstay of treatment for many years. Its action is essentially bacteriostatic and long-term continuous daily dosage was necessary when it was used alone. Because resistant strains of *M. leprae* are now widespread, this simple approach to treatment can no longer be recommended. Combination therapy has become essential to inhibit the emergence of resistance. Rifampicin and clofazimine, which are relatively expensive, are most frequently used with dapsone in this way. A thioamide (ethionamide or protionamide) is sometimes used

Recommended treatment regimens for leprosy[a]

Drug	Paucibacillary leprosy[b]	Multibacillary leprosy[c]
Dapsone	100 mg daily	100 mg daily
Rifampicin	600 mg monthly (supervised)	600 mg monthly (supervised)
Clofazimine		50 mg daily + 300 mg monthly (supervised)

[a] Doses are for adults of 50–70 kg.
[b] Minimum duration of treatment: 6 months.
[c] Minimum duration of treatment: 2 years and, when monitoring is feasible, until skin smears are negative.

instead of clofazimine, but close medical supervision and monitoring of liver function are then required since these drugs are markedly hepatotoxic. Minocycline and some fluoroquinolones have shown promising bactericidal effects against *M. leprae* in mice but data from formal trials in humans are still awaited.

Relatively short courses of oral multidrug therapy are currently recommended by WHO [1] on the assumption that the combined drugs will kill any single-drug-resistant strains initially present and will prevent their subsequent emergence during the period of treatment (see table above). It has been shown that compliance with the shorter-term treatment is better than with the older, longer-term regimens and it is expected that, with the current widespread adoption of the combined regimens by national control programmes, the intensity of transmission of *M. leprae* will be reduced.

After treatment, patients with paucibacillary disease should remain under supervision for at least a further 2 years and those with multibacillary leprosy for a further 5 years.

Management of exacerbations

Chemotherapy should never be suspended during an exacerbation. Acute neuritis must be treated as a medical emergency. Type I reactions frequently respond to corticosteroids administered immediately in high dosage (for adults, 40–60 mg of prednisolone daily) for several days and subsequently in gradu-

---

[1] *A guide to leprosy control*, 2nd ed. Geneva, World Health Organization, 1988, and WHO Technical Report Series, No. 768, 1988 (*WHO Expert Committee on Leprosy: sixth report*).

ally reduced dosage over a period of several weeks or months in accordance with the clinical response. Surgery may be needed to relieve focal compression of peripheral nerves. The pain and inflammation associated with type II reactions (erythema nodosum leprosum) may be largely suppressed by analgesics and anti-inflammatory agents. Thalidomide, which is a potent anti-inflammatory agent, can offer valuable relief, but should not be used in women of childbearing age. The initial adult dosage of 100–400 mg each evening, which may cause drowsiness, should be reduced gradually. Corticosteroids and clofazimine are alternatives.

Care of ocular lesions

The eyes should be examined regularly even in patients with no ocular symptoms. Closure of the eyelids can often be improved in mild degrees of lagophthalmos simply by application of a good quality dermal tape, but in severe cases tarsorrhaphy is necessary. In the absence of corneal ulcers, iridocyclitis should be treated with topical corticosteroids applied six times daily and with daily instillations of atropine or another long-acting mydriatic until the attack subsides. Corticosteroid application should then be gradually reduced in frequency over a week, and continued twice daily for at least one further week before final withdrawal. Mydriatics should similarly be administered two or three times weekly for 2–4 weeks after the initial attack has subsided.

Corneal abrasions need to be treated as early as possible with an antibiotic eye ointment (tetracycline 1%). This is particularly important in patients with lagophthalmos and when corneal sensitivity is impaired. Several applications daily may be needed for a prolonged period. The patient should be referred urgently to an ophthalmologist if a corneal ulcer develops.

Care of trophic lesions

For details of the initial care of trophic lesions, see *A guide to leprosy control*, 2nd ed. (Geneva, World Health Organization, 1988). Secondary bacterial infection of trophic ulcers may lead to osteomyelitis, necessitating antibiotic therapy and surgical care.

Clofazimine

Group: antimycobacterial agent
Capsule 50 mg, 100 mg

General information

A substance with both antileprosy and anti-inflammatory activity. It is weakly bactericidal against M. *leprae* and antimicrobial activity can be demonstrated in humans only after continuous exposure for about 50 days. When taken orally it is well absorbed and intermittent dosage is effective because the drug accumulates in fatty tissues and the cells of the reticuloendothelial system. It is very slowly eliminated in the faeces with a half-life of about 70 days. As yet, resistance to clofazimine is rare.

Clinical information

Uses
Treatment of:
- multibacillary leprosy in combination with dapsone and rifampicin
- type II reactions, as an alternative or in addition to corticosteroids or thalidomide.

Dosage and administration
Clofazimine should be taken with food or milk.

Multibacillary leprosy
Adults: 50 mg daily for at least 2 years, supplemented by one monthly supervised dose of 300 mg.

Children: 50 mg given on alternate days for at least 2 years, supplemented by one monthly supervised dose of 200 mg.

Erythema nodosum leprosum
Adults and children: 200–300 mg daily (given under medical supervision) for no longer than 3 months.

Precautions
Patients with pre-existing gastrointestinal disease should be kept under medical supervision. If symptoms become severe, it may be necessary to reduce the dosage or to prolong the interval between doses. Liver function and creatinine clearance should be monitored.

Use in pregnancy
Since leprosy is exacerbated during pregnancy, it is important that treatment should be continued. Infants exposed *in utero* may be more deeply pigmented than normal at birth.

Adverse effects
Reversible skin discoloration may occur during treatment to an extent that some lighter-skinned patients find unacceptable. Discoloration of the hair, cornea, conjunctiva, tears, sweat, sputum, faeces and urine also occurs.

Dose-related gastrointestinal symptoms include pain, nausea, vomiting and diarrhoea.

Clofazimine tends to accumulate in the phagocytic monocytes of the small intestine. Prolonged treatment with doses higher than those currently recommended for the treatment of multibacillary disease has resulted in mucosal and submucosal oedema severe enough to produce symptoms of subacute small-bowel obstruction. Because of this rare but serious adverse effect it is recommended that the high dosages used in the treatment of erythema nodosum leprosum should be given only under medical supervision and for no longer than 3 months.

Storage
Clofazimine capsules should be kept in well-closed containers.

Dapsone

Group: antileprosy agent
Tablet 50 mg, 100 mg

General information

A sulfone that remains of prime importance in the treatment of leprosy. Dapsone is both bacteriostatic and weakly bactericidal against *M. leprae*, the minimum inhibitory concentration for fully sensitive organisms being approximately 0.003 micrograms/ml. However, resistant strains can develop *de novo* during prolonged treatment with dapsone alone, and their incidence is increasing in previously untreated patients. In some areas the prevalence of primary resistance is currently estimated to be as high as 40%.

After absorption from the gastrointestinal tract, dapsone is distributed widely in body tissues and it is subsequently retained selectively in skin, muscle, liver and kidneys. It is partially acetylated or conjugated in the liver and ultimately excreted in the urine as metabolites. A dose of 100 mg produces a peak serum concentration of approximately 2 micrograms/ml, which declines with a half-life of 1–2 days.

Clinical information

Uses
Treatment of paucibacillary and multibacillary leprosy in combination with other antileprosy drugs.

Dosage and administration

Paucibacillary leprosy (in combination with rifampicin)
Adults and children: 1–2 mg/kg daily (adult dose usually 100 mg) for at least 6 months.

Multibacillary leprosy (in combination with rifampicin and clofazimine)
Adults and children: 1–2 mg/kg daily (adult dose usually 100 mg) for at least 2 years.

Contraindications
- Known hypersensitivity to sulfones.
- Severe anaemia.

Precautions
Pre-existing severe anaemia should be treated before dapsone therapy is started.

Dapsone can induce haemolysis of varying degree, particularly in patients with glucose-6-phosphate dehydrogenase deficiency, and dose-dependent methaemoglobinaemia may supervene during the second week of treatment. The clinical response and the blood count must therefore be closely monitored in susceptible patients during the first weeks of treatment. Dapsone therapy should not be discontinued if exacerbations occur.

Use in pregnancy
Since leprosy is exacerbated during pregnancy, it is important that treatment should be continued.

Adverse effects
Dapsone is generally well tolerated at recommended dosages, but symptoms of gastrointestinal irritation occasionally occur. Other, less common reactions include headache, nervousness and insomnia.

Blurred vision, paraesthesiae, reversible peripheral neuropathy, drug fever, skin rashes and psychoses have also been reported. Hepatitis, Herxheimer reactions and agranulocytosis may rarely occur.

Dapsone (continued)

Overdosage
Acute overdosage results in nausea, vomiting and hyperexcitability. Orally administered activated charcoal may enhance the elimination of dapsone.

Storage
Dapsone tablets should be kept in well-closed containers protected from light.

Rifampicin

Group: antimycobacterial agent
Capsule or tablet 150 mg, 300 mg

General information
A semisynthetic derivative of rifamycin, a complex macrocyclic antibiotic that inhibits ribonucleic acid synthesis in a broad range of microbial pathogens.

Rifampicin is lipid-soluble. Following oral administration, it is rapidly absorbed and distributed throughout the cellular tissues and body fluids; if the meninges are inflamed, significant amounts enter the cerebrospinal fluid. A single dose of 600 mg produces a peak serum concentration of about 10 micrograms/ml in 2–4 hours, which subsequently decays with a half-life of 2–3 hours. It is extensively recycled in the enterohepatic circulation, and metabolites formed by deacetylation in the liver are eventually excreted in the faeces.

Since resistance readily develops, rifampicin must always be administered in combination with other effective antimycobacterial agents.

Clinical information

Uses
Treatment of paucibacillary and multibacillary leprosy in combination with other antileprosy drugs.

Dosage and administration
Administration of rifampicin should always be supervised. The drug should preferably be given at least 30 minutes before meals, since absorption is reduced when it is taken with food.

Paucibacillary leprosy (in combination with dapsone)
Adults: 600 mg once a month for at least 6 months.

Children: 10 mg/kg once a month for at least 6 months.

Multibacillary leprosy (in combination with dapsone and clofazimine)
Adults: 600 mg once a month for at least 2 years.

Children: 10 mg/kg once a month for at least 2 years.

Contraindications
- Known hypersensitivity to rifamycins.
- Hepatic dysfunction.

Precautions
Serious immunological reactions resulting in renal impairment, haemolysis or thrombocytopenia are on record in patients who resume taking rifampicin after a prolonged lapse of treatment. In this rare situation it should be immediately and definitively withdrawn.

Careful monitoring of liver function is required in the elderly and in patients who are alcohol-dependent or have hepatic disease.

Patients should be warned that treatment may produce a reddish discoloration of urine, tears and saliva and that contact lenses may be irreversibly stained.

Use in pregnancy
Since leprosy is exacerbated during pregnancy it is important that treatment should be continued.

Vitamin K should be administered to the infant at birth because of the risk of postnatal haemorrhage.

Adverse effects
Rifampicin is well tolerated by most patients at currently recommended dosages, although gastrointestinal intolerance can be unacceptably severe. Other adverse effects (rashes, fever, influenza-like syndrome and thrombocytopenia) are more likely to occur during intermittent (three times weekly) administration, as sometimes used in the treatment of tuberculosis. Temporary oliguria, dyspnoea and haemolytic anaemia have also occasionally been reported in these circumstances, but they have never been associated with the monthly dosage schedules advocated in leprosy.

Moderate rises in serum concentrations of bilirubin and transaminases, which are common at the outset of treatment, are often transient and without clinical significance. However dose-related hepatitis can occur, which is potentially fatal. It is consequently important not to exceed the maximum recommended daily dose of 10 mg/kg (600 mg).

Drug interactions
Rifampicin is a potent inducer of hepatic enzymes when it is administered daily. The dosage of other drugs metabolized in the liver may need to be increased when they are taken concomitantly. These include corticosteroids, steroid contraceptives, oral hypoglycaemic agents, oral anticoagulants, phenytoin, cimetidine, quinidine, ciclosporin and digitalis glycosides. However, this effect is less pronounced when rifampicin is administered monthly.

Because enzyme induction reduces the reliability of steroid contraceptives, patients should be strongly advised to use a nonhormonal method of birth control throughout treatment and for at least 1 month subsequently.

Biliary excretion of radiocontrast media and sulfobromophthalein sodium may be reduced and microbiological assays for folic acid and vitamin B_{12} disturbed.

Overdosage
Gastric lavage may be of value if undertaken within a few hours of ingestion. Very large doses may depress central nervous function. There is no specific antidote and treatment is supportive.

Storage
Capsules and tablets should be kept in tightly closed containers, protected from light.

Ethionamide and protionamide

Group: antileprosy agent
Tablet 125 mg, 250 mg

General information
The thioamides ethionamide and protionamide are derivatives of thioisonicotinic acid. Both are weakly bactericidal to *M. leprae*. Their biological properties and therapeutic potency are very similar.

Ethionamide & protionamide (cont.)

They are readily absorbed from the gastrointestinal tract and widely distributed throughout body tissues. The plasma half-life of both compounds is approximately 2–4 hours and they are excreted in the urine largely as metabolites.

Clinical information

Uses
Treatment of multibacillary leprosy (in combination with dapsone and rifampicin), to prevent emergence of drug resistance. Because of their hepatotoxicity thioamides should be used only when clofazimine is unacceptable or not available.

Dosage and administration
Adults and children: 5.0 mg/kg daily (usual adult dose 250–375 mg).

Contraindications
- Known hypersensitivity.
- Hepatic dysfunction.
- Pregnancy.

Precautions
Liver function tests must be performed at the start of treatment and repeated periodically throughout.

Adverse effects
Liver dysfunction and toxic hepatitis may occur.

Gastrointestinal disturbances are common. Other reported adverse effects include acne, allergic reactions, alopecia, convulsions, dermatitis, diplopia, dizziness, headache, hypotension, peripheral neuropathy and rheumatic pains.

Storage
Ethionamide and protionamide tablets should be kept in tightly closed containers protected from light.

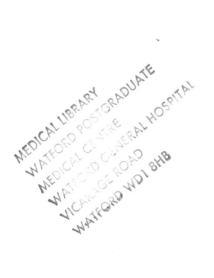

Nonspecific mycobacterial diseases

Nontuberculous mycobacteria are ubiquitous in the environment. They exist in food, soil and water, on the surface of many plants, and in buildings, particularly within water pipes. For many years they were thought to be implicated in human disease only as saprophytic contaminants in tuberculous lesions. Now, several species are recognized to be facultative parasites capable of causing chronic granulomatous diseases that can be pathologically indistinguishable from tuberculosis.

These infections are not readily identified because the causative bacteria can be distinguished from *M. tuberculosis* only by specialized methods. However, they have latterly attracted attention for two reasons. Firstly, wherever tuberculosis has declined within the population at large, they now account for a greater proportion of cases of granulomatous disease. Secondly, like tuberculosis, they have emerged as common secondary infections among patients with acquired immunodeficiency syndrome (AIDS).

Most localized infections result from inoculation of organisms into the skin. Pulmonary infection usually occurs only in patients with predisposing disease, while disseminated infection is confined almost exclusively to patients with impaired immune responses. Thus far, there is no evidence of case-to-case transmission.

Clinically, four types of disease are described:

Localized cutaneous lesions
Inoculation of organisms into superficial abrasions and into puncture wounds can result in the formation of localized nodular or ulcerative lesions. The organisms most commonly implicated are *M. marinum*, which colonizes swimming pools and fish aquaria, and *M. ulcerans*, which is largely restricted to Australia and some tropical regions and causes deep necrotic lesions known as Buruli ulcers. *M. haemophilum* has more recently been associated with similar lesions in immunosuppressed patients. Abscesses resulting from contaminated injections have most frequently been attributed to two rapidly growing species,

M. chelonei and *M. fortuitum*. More such cases are to be anticipated among drug addicts who are immunosuppressed as a result of AIDS but, as yet, most have occurred either among diabetics or following the injection of contaminated drugs and vaccines.

Pulmonary disease
The lung is the most frequent site of opportunistic mycobacterial infection and the lesions are clinically and radiologically indistinguishable from pulmonary tuberculosis. Predisposing conditions include chronic bronchitis, occupational dust-induced diseases, residual tuberculous lesions, cystic fibrosis, carcinoma, AIDS and other causes of immunosuppression. Most recorded cases have been attributed to the *M. avium–intracellulare* complex, *M. kansasii* and, to a lesser extent, *M. xenopi*, but in some regions *M. scrofulaceum*, *M. chelonei*, *M. szulgai* and *M. malmoense* are also of significance. Any of the organisms implicated in pulmonary disease can also cause lymphadenitis and other forms of nonpulmonary disease more commonly associated with *M. tuberculosis*.

Lymphadenitis
The lesions are usually unilateral and self-limiting and most cases occur in children under 5 years old. However, lymphadenitis is also sometimes a prominent feature of disseminated disease in adults. Most reported cases have been attributed to the closely related *M. avium–intracellulare* complex and *M. scrofulaceum* (known as the MAIS complex).

Disseminated disease
Single or multiple foci of granulomatous disease can occur in virtually any system or organ and, when cellular immunity is depressed, dissemination of the infection can occur as rapidly as in miliary tuberculosis. Most such cases have been attributed to the *M. avium–intracellulare* complex and to *M. chelonei*.

Management
Diagnosis is dependent upon the clinical characteristics of the disease and identification of the causative organism, when this is possible. Although all mycobacterial infections are presumed to give rise to a positive result in the tuberculin test, seroconversion is most likely to have resulted from previous infection with

M. tuberculosis. The test is thus of little, if any, practical value in the diagnosis of nontuberculous infections.

It is not known whether BCG vaccines protect against infection by any of the nontuberculous mycobacteria and, as yet, no specific vaccines have been developed. The management of established infection is determined by the anatomical focus of the disease, the identity of the organism, the age of the patient, and the competence of the immune system.

General principles

Deep and widely disseminated infections can be treated only by chemotherapy. However, the response, even when treatment is prolonged, is uncertain, and surgical resection—now rarely employed in tuberculosis—remains of value in localized nontuberculous pulmonary disease. Surgical excision is also frequently used to hasten resolution of localized lymphadenopathy and of solitary skin lesions, even though these lesions are likely to be self-limiting.

The isolation of nontuberculous mycobacteria from biopsy of a chronic granulomatous lesion is generally evidence of a causal association. However, the identification of these ubiquitous facultative parasites in sputum or urine requires guarded interpretation as their presence does not necessarily imply a pathogenic role. Only when tuberculosis has been rigorously excluded and positive cultures have been consistently obtained over a period of several weeks should the patient be committed to the prolonged, costly and sometimes hazardous courses of chemotherapy required. Whenever possible, the identity of the causative organism and its sensitivity to candidate antibiotics should be established within a specialized reference laboratory. However, *in vitro* sensitivity tests can be misleading, and treatment may need to be determined empirically on the basis of published case reports and retrospective surveys.

Selection of chemotherapeutic agents

Most experience has been gained in the treatment of localized pulmonary disease caused by the more prevalent and relatively slowly growing *M. kansasii*, *M. malmoense* and, in immunocompetent hosts, the *M. avium–intracellulare* complex. These species are usually ultimately responsive to standard antituberculosis chemotherapy—even though the *M. avium–intracellulare* complex can be relatively resistant *in vitro*—but it is often necessary to administer a combination of rifampicin, ethambutol and isoniazid for at least 18 to 24 months.

Several other antibiotics not normally used in antituberculosis therapy have been claimed to be of value. These include erythromycin in infections due to *M. kansasii, M. scrofulaceum* and, less reliably, the *M. avium–intracellulare* complex, and the combination of sulfamethoxazole and trimethoprim in infections attributed to the *M. avium–intracellulare* complex, *M. chelonei, M. marinum* and *M. xenopi*. Reports also exist of *M. chelonei* and *M. fortuitum* infections responding to a combination of amikacin and doxycycline, and of *M. marinum* infections responding to minocycline. However, the evidence supporting the use of these drugs is largely anecdotal, and firm recommendations cannot be made.

Clofazimine, which is concentrated in fatty tissues and in cells of the reticuloendothelial system, is claimed to be of particular value in the suppression of disseminated disease due to the *M. avium–intracellulare* complex. It has been advocated in combination with rifamycin in the management of opportunistic infections in patients with AIDS, but insufficient information is currently available to determine whether this regimen has significant effect on morbidity and survival time.

Rifampicin

Group: antimycobacterial agent
Capsule or tablet 150 mg, 300 mg

General information

A semisynthetic derivative of rifamycin, a complex macrocyclic antibiotic that inhibits ribonucleic acid synthesis in a broad range of microbial pathogens.

Rifampicin is lipid-soluble. Following oral administration, it is rapidly absorbed and distributed throughout the cellular tissues and body fluids; if the meninges are inflamed, significant amounts enter the cerebrospinal fluid. A single dose of 600 mg produces a peak serum concentration of about 10 micrograms/ml in 2–4 hours, which subsequently decays with a half-life of 2–3 hours. It is extensively recycled in the enterohepatic circulation, and metabolites formed by deacetylation in the liver are eventually excreted in the faeces.

Since resistance readily develops, rifampicin must always be administered in combination with other effective antimycobacterial agents.

Clinical information

Uses
In combination with ethambutol and isoniazid in the treatment of infections due to *M. kansasii*, *M. malmoense*, *M. xenopi* and, in immunocompetent hosts, the *M. avium–intracellulare* complex.

Dosage and administration
Rifampicin should preferably be given at least 30 minutes before meals, since absorption is reduced when it is taken with food.

Adults and children: 10 mg/kg (maximum 600 mg) daily or three times weekly for 2 years. There is some evidence to suggest that *M. kansasii* infections require treatment for only 1 year.

Contraindications
- Known hypersensitivity to rifamycins.
- Hepatic dysfunction.

Precautions
Serious immunological reactions resulting in renal impairment, haemolysis or thrombocytopenia are on record in patients who resume taking rifampicin after a prolonged lapse of treatment. In this rare situation it should be immediately and definitively withdrawn.

Careful monitoring of liver function is required in the elderly and in patients who are alcohol-dependent or have hepatic disease.

Patients should be warned that treatment may produce reddish discoloration of urine, tears, saliva and sputum, and that contact lenses may be irreversibly stained.

Use in pregnancy
Treatment should not be interrupted or postponed during pregnancy.

Vitamin K should be administered to the infant at birth because of the risk of postnatal haemorrhage.

Adverse effects
Rifampicin is well tolerated by most patients at currently recommended doses, although gastrointestinal intolerance can be unacceptably severe. Other adverse effects (skin rashes, fever, influenza-like syndrome and thrombocytopenia) are more likely to occur with intermittent than with daily administration. Temporary oliguria, dyspnoea and haemolytic anaemia have also been reported. These

Rifampicin (continued)

reactions subside when daily dosage is instituted.

Moderate rises in serum concentrations of bilirubin and transaminases, which are common at the outset of treatment, are often transient and without clinical significance. However, dose-related hepatitis can occur, which is potentially fatal. It is consequently important not to exceed the maximum recommended daily dose of 10 mg/kg (600 mg).

Drug interactions

Rifampicin induces hepatic enzymes, and may increase the dosage requirements of drugs metabolized in the liver. These include corticosteroids, steroid contraceptives, oral hypoglycaemic agents, oral anticoagulants, phenytoin, cimetidine, quinidine, ciclosporin and digitalis glycosides.

Patients should consequently be advised to use a nonhormonal method of birth control throughout treatment and for at least 1 month subsequently.

Biliary excretion of radiocontrast media and sulfobromophthalein sodium may be reduced and microbiological assays for folic acid and vitamin B_{12} disturbed.

Overdosage

Gastric lavage may be of value if undertaken within a few hours of ingestion. Very large doses may depress central nervous function. There is no specific antidote and treatment is supportive.

Storage

Capsules and tablets should be kept in tightly closed containers, protected from light.

Ethambutol

Group: antimycobacterial agent
Tablet 100–400 mg (hydrochloride)

General information

A synthetic congener of 1,2-ethanediamine that is bactericidal against some nonspecific mycobacteria.

It is readily absorbed from the gastrointestinal tract. Plasma concentrations peak in 2–4 hours and decay with a half-life of 3–4 hours. Ethambutol is excreted in the urine both unchanged and as inactive hepatic metabolites. About 20% is excreted in the faeces as unchanged drug.

Clinical information

Uses

In combination with rifampicin and isoniazid in the treatment of infections due to *M. kansasii, M. malmoense, M. xenopi* and, in immunocompetent hosts, the *M. avium–intracellulare* complex.

Dosage and administration

Adults: 25 mg/kg daily for no more than 2 months followed by 15 mg/kg daily; or 40 mg/kg three times weekly for 2 years.

Children: 15 mg/kg daily for 2 years.

There is some evidence to suggest that *M. kansasii* infections require treatment for only 1 year.

Dosage must always be carefully calculated on a weight basis to avoid toxicity, and should be reduced in patients with impaired renal function.

Contraindications
- Known hypersensitivity.
- Pre-existing optic neuritis from any cause.
- Inability (for example due to young age) to report symptomatic visual disturbances.
- Creatinine clearance of less than 50 ml/minute.

Precautions
Patients should be advised to discontinue treatment immediately and to report to a doctor should their sight or perception of colour deteriorate. Patients who are too young or who are otherwise unable to comprehend this warning should not receive ethambutol.

Whenever possible, renal function should be assessed before treatment.

Use in pregnancy
Treatment should not be interrupted or postponed during pregnancy.

Adverse effects
Dose-dependent optic neuritis can readily result in impairment of visual acuity and colour vision. Early changes are usually reversible, but blindness can occur if treatment is not discontinued promptly.

Signs of peripheral neuritis occasionally develop in the legs.

Overdosage
Emesis and gastric lavage may be of value if undertaken within a few hours of ingestion. Subsequently, dialysis may be of value. There is no specific antidote and treatment is supportive.

Storage
Tablets should be stored in well-closed containers.

Isoniazid

Group: antimycobacterial agent
Tablet 100–300 mg
Injection 25 mg/ml in 2-ml ampoule

General information
Isoniazid, the hydrazide of isonicotinic acid, is bactericidal against some nonspecific mycobacteria.

It is rapidly absorbed and diffuses readily into all fluids and tissues. The plasma half-life, which is genetically determined, varies from less than 1 hour in fast acetylators to more than 3 hours in slow acetylators. It is largely excreted in the urine within 24 hours, mostly as inactive metabolites.

Clinical information
Uses
In combination with rifampicin and isoniazid in the treatment of infections due to *M. kansasii, M. malmoense, M. xenopi* and, in immunocompetent hosts, the *M. avium–intracellulare* complex.

Dosage and administration
Isoniazid is normally taken orally but it may be administered intramuscularly to critically ill patients.

Isoniazid (continued)

Adults and children: 5 mg/kg daily or 15 mg/kg three times weekly for 2 years. There is some evidence to suggest that *M. kansasii* infections require treatment for only 1 year.

Contraindications
- Known hypersensitivity.
- Active hepatic disease.

Precautions
Serum concentrations of hepatic transaminases should be monitored whenever possible.

Patients at risk of peripheral neuropathy as a result of malnutrition, chronic alcohol dependence or diabetes should additionally receive pyridoxine, 10 mg daily. Where the standard of health in the community is low this should be offered routinely.

Epilepsy should be effectively controlled since isoniazid may provoke attacks.

Use in pregnancy
Treatment should not be interrupted or postponed during pregnancy.

Adverse effects
Isoniazid is generally well tolerated at recommended doses. Systemic or cutaneous hypersensitivity reactions occasionally occur during the first weeks of treatment.

The risk of peripheral neuropathy is excluded if vulnerable patients routinely receive supplements of pyridoxine. Other less common forms of neurological disturbance, including optic neuritis, toxic psychosis and generalized convulsions, can develop in susceptible individuals, particularly in the later stages of treatment, and occasionally necessitate the withdrawal of isoniazid.

Hepatitis is an uncommon but potentially serious reaction that can usually be averted by prompt withdrawal of treatment. More often, however, a sharp rise in serum concentrations of hepatic transaminases at the outset of treatment is not of clinical significance. If the enzyme levels drop rapidly when dosage is suspended, they are unlikely to rise sharply again when treatment is resumed.

Drug interactions
Isoniazid tends to raise plasma concentrations of phenytoin and carbamazepine by inhibiting their metabolism in the liver. The absorption of isoniazid is impaired by aluminium hydroxide.

Overdosage
Nausea, vomiting, dizziness, blurred vision and slurring of speech occur within 30 minutes to 3 hours of overdosage. Massive poisoning results in coma preceded by respiratory depression and stupor. Severe intractable seizures may occur. Emesis and gastric lavage can be of value if instituted within a few hours of ingestion. Subsequently, haemodialysis may be of value. Administration of high doses of pyridoxine is necessary to prevent peripheral neuritis.

Storage
Tablets should be kept in well-closed containers, protected from light. Solution for injection should be stored in ampoules protected from light.